Out of the way world here comes humanity!

BY KEITH HILL

POETRY

The Ecstasy of Cabeza de Vaca
The Lounging Lizard Poet of the Floating World
The Bhagavad Gita: A new poetic translation
Interpretations of Desire:
Mystical love poems by the Sufi master Ibn 'Arabi
I Cannot Live Without You:
Selected poetry of Mirabai and Kabir
Psalms of Exile and Return

FICTION

Blue Kisses
Puck of the Starways

NON-FICTION

The New Mysticism
The God Revolution
Striving To Be Human

Out of the way world
Here comes humanity!

poems / antipoems

Keith Hill

DISJUNCT BOOKS

Published in 2022 by Disjunct Books
an imprint of Attar Books, New Zealand

Paperback ISBN 978-1-99-115707-2
Hardcover ISBN 978-1-99-115708-9
Ebook ISBN 978-1-99-115709-6

Copyright © Keith Hill 2022

Keith Hill's right to be identified as author of this work is asserted in accordance with Section 96 of the Copyright Act 1996.

All rights reserved. Except for fair dealing or brief passages quoted in a newspaper, magazine, radio, television or internet review, no part of this book may be reproduced in any form or by any means, or in any form of binding or cover other than that in which it is published, without permission in writing from the Publisher. This same condition is imposed on any subsequent purchaser.

Disjunct Books is published by Attar Books, a New Zealand publisher which focuses on work that explores today's spiritual experiences, culture, concepts and practices. For more information visit our website:

www.attarbooks.com
www.keithhillauthor.org

Contents

A word to the young   9

Meet the neighbours

The politician	15
The troll	16
The worker	17
The millennial	18
The curmudgeon	19
The anti-vaxer	20
The voter	21
The patriot	22
The gamers	23
The world	24

This is the news

Supercut: A season of Covid-19	27

Boogying with the virus

A modest proposal	57
We need to stop dithering	60
Hooray for the workers	64
Psalm for the End Times	70
Let's give it another go	78

The world needs therapy

The individual's soliloquy	87
America	93
Looky yonder	99
How to found a nation	102
Request	111
Acknowledgements	115
About the author	117
To the reader	118

To the reader

We are on a collision course with our future. In these pages I have attempted to capture our uneasy dance around this fact. While our unease certainly preexisted Covid-19, the virus has made visible fault lines—driven by self-interest, disinformation and denial—we have long ignored while caught up in the head-rush activity that, over-simplistically, we call human progress. Not that I am against progress, but the time has come to acknowledge that in our relentless grasping for "more" we have not been accounting for the "less" we will leave those living on this planet after we have gone.

A situation this serious requires straight talking. I have been inspired by the straight talking of Lawrence Ferlinghetti, his democratic dog, and his carnivalesque mind; Allen Ginsberg, who while putting his queer shoulder to the wheel saw the world is a baby crawling towards a death chamber; and Nicanor Parra, who perfected the art of looking life in the eye while whistling from the bottom of a coffin.

Their work makes clear that straight talking cannot be practised naively. Each writer needs their own strategy for talking about and to the world. I have chosen the satiric mode, in acknowledgement of the fact that, more than information or knowledge, what people most want today is to be right.

I have no illusions that you will agree with all, or any, of what follows. But I do have one request. As you read consider this: He waka eke noa—we are in this together. Otherwise, and in accord with current social settings, it is open slather.

A word to the young

Let's start with some friendly advice.
It's offered with the best intentions
but you may wish to follow your
elders' lead and consider nothing
this serious in a state of sobriety.
The truth is your elders hate you.
They have written promissory notes
for the future and crossed their
fingers they will be long gone
when nature comes to collect.
And that's the good news.

You have surely noticed by now
your elders' approach to solving
today's most intractable problems
is to employ experts who closely
analyse all relevant data then blame
those they won't allow to respond.
The blamed when I was young included
hippies students communists
unions workers the unemployed
radicals liberals immigrants
unmarried mothers abortionists the PC
feminists homosexuals transbetweeners
out of touch old people and the young.
Having identified the requisite causes
the experts then write up their findings
in bulging reports that contain copious
data analysis graphs and footnotes

the principal purpose of which is
to show value for money.
With suitable ceremony these reports
are presented to the authorities
who skim read the executive summaries
then recite with great fervour
a speech recycled from the year before
noting the need for "all of us together"
to energetically address X Y or Z.
Months later a media release put out
late one Friday night states the report
has been filed in a cabinet labelled
For Future Consideration.

This is how the modern world advances—
in a stuttering two-step performed
to music played by an om-pah-pah band
marching in a gleaming town hall
where if you look behind the scenes
you'll see the walls are constructed
from imported plebiscites propped up
by poles extruded from hydrocarbons
precariously tied with contested agendas.
The impressive gleam is produced by
lacquer blended from ambition dissolved
in pragmatism painted on by
minimally paid migrant workers
who are blamed both for being
and for not being in the country
but business leaders in private admit
they need because cheap labour
drives their profit margins.

The conclusion is we should be grateful
because things could be worse.

As your stint in the education machine
comes to a close you are likely wondering
what your own future promises.
From my educated perspective
gained by having screwed up my own life
several times
I have just three pieces of advice.
Don't get distracted by what's in front of you.
(The big decisions are happening elsewhere.)
Don't be naive.
(When have words and actions ever aligned?)
And don't give in to despair.
Because things could be worse.
(Actually they soon will be.)
I guess that's it.
Otherwise chin up stay positive don't slouch.
And don't make any long-term plans.

Meet the neighbours

The politician

The politician scratches the audience's back with
their favourite slogans left over from the last election.

The politician is a centrist who hangs his right shoe
on his left ear and speaks out the side of his mouth.

The politician first denies a problem exists then blames
the opposition then wants it left for the next election.

The politician ignores the students outside chanting
"Dinosaur!" as he studies the latest polling numbers.

The politician is gob-smacked when told his policies
align exactly with the prejudices of voters who back him.

The politician blows up indignantly when he's accused
of wanting to win more than do what the country needs.

The politician projects himself as the last great hope
in a world where hope has replaced having a future.

The troll

The troll walks the streets with his outrage trotting
on a leash beside him stopping to let it pee and poop.

The troll refuses to carry a pooper scooper because
that would infringe his right to free expression.

The troll's outrage is the emotional support animal
he needs to get through each bullshit day.

The troll lets his outrage off the leash because
it's inhumane to always keep an animal restrained.

The troll grins as his outrage rampages through the park
biting whoever looks different and scaring everyone else.

The troll laughs as free speech advocates defend his
right to let his outrage savage whoever he can't stand.

The troll walks off leaving behind a line of steaming poop
to signpost his contribution to the country's freedom.

The worker

The worker doesn't understand how it is that
the freer the market gets the more everything costs.

The worker gloomily calculates the cost of daily
existence on the backs of rent food and utilities bills.

The worker is in the debit column of an economy
where being in the majority doesn't add up.

The worker uses a dart board to select the family's
aspirational goals hopes expectations and dreams.

The worker is resigned to their family's health depending
on whether the boss boosts their pay to a livable wage.

The worker can't decide what's worse: foreign machines
taking their job or the 1% taking their kids' future.

The worker is the engine of the economy until they're
laid off when they become ungrateful bludgers and scabs.

The millennial

The millennial wakes each morning to stare glumly
into a hole that was once the promise of a life.

The millennial orders a double shot decaf almond
milk latte and hunts for prophecies in the froth.

The millennial seeks solace in spirituality because their
parents' neoliberal capitalism is crashing the planet.

The millennial zooms seminars and masterclasses
hoping to meet the person they would like to be.

The millennial coats themself in probiotic goop
and surrenders to the process of transformation.

The millennial tries to beat depressing emptiness
by learning the Tao of mindfully being without having.

The millennial has a satori they'll never be trapped by
possessions because boomers have taken everything.

The curmudgeon

The curmudgeon visits a neighbourhood market
and overhears others call him "a difficult bastard".

The curmudgeon decides he needs to be less acerbic
and more accommodating so makes an effort to fit in.

The curmudgeon realises that means all the stupidity he
sees he'll just have to let slide like water off a duck's back.

The curmudgeon wades into a public debate on e-bikes
replacing cars and the need to vary transport corridors.

The curmudgeon wonders how ducks can swim through
so much garbled ignorance without drowning in gunk.

The curmudgeon takes a deep breath counts to a million
then writes to the editor proclaiming they're all wrong.

The curmudgeon sits back relieved knowing where he
fits in is on the outer where truth and the bogeyman live.

The anti-vaxer

The anti-vaxer has her hair done before she goes live on
Instagram to declare Covid-19 a government plandemic.

The anti-vaxer is a WAG who feels obliged to repay for her
privileged position by sharing the truth with the world.

The anti-vaxer was once blind but turned her life round
via YouTube and doing the hard yards in the gym.

The anti-vaxer is anti-mask because it's a form of slavery
that takes freedom of choice and is hell on her lipstick.

The anti-vaxer connects to the world through her phone
but is terrified 5G towers will infect her kids with Covid.

The anti-vaxer knows evil forces manipulate the world
but experience tells her you can't lead a sheep to water.

The anti-vaxer gets trolled for her beliefs but deep
in her heart all she wants is for everyone to be saved.

The voter

The voter exercises their right to transcend their self-
interest and back the party giving tax breaks to the rich.

The voter decides wealthy businessmen make the best
leaders because they've prospered in a brutal economy.

The voter knows for the nation to prosper we have
to stop listening to whiners panderers and muppets.

The voter hates socialist sharing because it's a dog-eat-
dog world they've struggled and there's no free lunch.

The voter wants hard-working innovators to be given
carte blanche to create the enriched hi-tech world we need.

The voter jeers at commie politicians who instead want
to dish out taxpayers' hard-earned dollars to lazy losers.

The voter despairs that the only time democracy isn't
a disaster is when those they voted for get elected.

The patriot

The patriot first doubted wondering how the aircraft
that hit the Pentagon could evaporate on impact.

The patriot's suspicions grew learning chemtrails lacing
the sky are designed to re-engineer humanity's genes.

The patriot concludes scientists' denials of this fact
proves how much governments control what we're told.

The patriot discovers globalisation is a ploy by a deep
state elite constructing a New World Order by stealth.

The patriot is shocked to learn the elite is riddled with
paedophiles who imprison children and drink their blood.

The patriot is horrified the elite is working with reptilian
ETs who spread Covid to enslave humanity via vaccines.

The patriot champions a cleansing storm to trump
the evil elite and put the good guys back in charge.

The gamers

Two gamers are so bored during lockdown they riff on game concepts to explore humanity's post-Covid future.

One gamer envisions creating a utopian game world with rules built on altruism and shared responsibility.

The other sees it as zero-sum with blinged-out terrain and a UX so intoxicating gamers can't stop playing.

One gamer pictures a Faustian avatar with knowledge and power who has to choose humanity's best course.

The other proposes a Faustian pact with Big Tech that climaxes in a death struggle between humans and AI.

Both agree they don't want a game world where fires hurricanes floods and pandemics decimate everything.

The gamers conceptually test all feasible options then decide the most realistic is to set their game on Mars.

The world

The world is a caffeinated collection of neuroses
storming from one burning teacup to the next.

The world is averse to opening the closed door
on the top of its head to let the light shine in.

The world has spent a century trying to persuade
itself modern civilisation is not a form of psychosis.

The world is overrun by macho snowflakes who
any moment will rant at you cancel you or shoot.

The world defensively argues life is an equal opport-
unity smorgasbord where we can eat all we can get.

The world struggles to process that its shadow
is prowling outside the house trying to break in.

The world gives up grabs the coat tails of the nearest
politician and is dragged screaming into the sunset.

This is the news

Supercut: A season of Covid-19

The following is edited from 33 news articles, research studies, blogs and opinion pieces written during 2020 and 2021. Collectively, these excerpts follow the arc of the pandemic and reflect on what faces us as we emerge from it. Sources are acknowledged at the end.

On 31 December 2019 Wuhan health officials
notify WHO of a case of "viral pneumonia".
On 9 January Chinese authorities declare it
a novel coronavirus and 2 days later send
its genetic sequences to WHO's scientists.
That same day the first death from the virus
is reported by Chinese media.
As WHO sends diagnostic advice worldwide
China's infections grow to 41 and Thailand
registers the first international case.
At this stage WHO doubts the virus
involves human-to-human transmission.
On 15 January Japan diagnoses the virus
in a person who had travelled to Wuhan—
one of China's largest transport hubs.
By 20 January 491 cases have been identified
in China—most in Wuhan—4 have died
and a WHO research team in Wuhan confirms
evidence of human-to-human transmission.
The US records its first case on 21 January
and France registers 4 cases 3 days later—
the latter had all travelled to Wuhan.

By month's end cases are multiplying across
China and infections have been diagnosed
in 29 countries in Asia Europe North America
Africa the Middle East and the Pacific.
On 30 January WHO's secretary-general
declares the outbreak a Public Health Emergency
of International Concern—its highest setting.
The first death outside China occurs 1 February.
The victim is a 44-year-old Chinese man
from Wuhan who dies in the Philippines.
On 11 February WHO announces that
the novel coronavirus will be called Covid-19.

"This is not nature's revenge,"
says leading US scientist Thomas Lovejoy.
"We did it to ourselves."
Inventor of the term biological diversity
Lovejoy claims the vast illegal wildlife trade
and humanity's excessive intrusion into nature
are to blame for the coronavirus pandemic.
"The wet markets of South Asia
and the bush meat markets of Africa—
it was just a matter of time before
something like this happened."

Minister Chen Wen has clarified facts about
China's fight against the Covid-19 outbreak.
Acknowledging Wuhan as the epicentre
Wen stated: "That's why China locked down
Wuhan city on 23 January 2020.
This clearly cut down the transmission
and is why we have very efficiently

very rigorously and effectively stopped
the virus spreading to other countries."
With regard to the safety of "wet markets"
there isn't such a concept in China.
"What China has is farmers' markets
where people can buy fresh fruit and meat.
A few sell live poultry but not wild animals.
This is banned by law."

From May 2017 to November 2019
monthly surveys were carried out in
Wuhan's pre-Covid food markets.
As an objective observer unconnected
to law enforcement X.X. was granted
unique and complete access to trading practices.
He collected data from 17 wet market shops
that sold live—and often wild—animals.
Wild animals are state property protected
by the Wild Animal Conservation Law.
Offenders face up to 15 years in prison.
Over the study period the 17 shops reported
total sales of 36,295 individuals
belonging to 38 wild animal species.
30% suffered wounds from gunshots or traps
implying illegal wild harvesting.
Almost all were sold alive in poor condition.
Most stores offered on-site butchering services.

> "The virus we're talking about
> a lot of people think it goes away
> as the heat comes in.
> We're in great shape though.

We have 12 cases 11 cases
and many are in good shape now.
I had a long talk with President Xi
2 nights ago. He feels very confident.
He feels that as I mentioned
by April or during April
the heat kills this kind of virus.
So that would be a good thing."

In mid-April $^1/_3$ of the world's population
is living under Covid-related restrictions.
Multiple countries have closed their borders
and ordered nationwide lockdowns—
global coronavirus cases pass 500,000—
official Covid deaths exceed 100,000—
New York's death toll passes 10,000—
Wimbledon cancels its tennis tournament—
Japan postpones the Summer Olympics
and declares a state of emergency—
the World Bank estimates the pandemic
could push 11 million into poverty—
UK PM Boris Johnson is in intensive care—
and news reports from Kentucky say a group
in their 20s had held a coronavirus party
after which at least 1 contracted Covid-19.
Governor Andy Beshear declares:
"We're battling for the health and even
lives of our parents and grandparents.
Don't be so callous as to intentionally
expose yourself to something that
can kill others—we have to be
much better than that."

Former Australian PM Tony Abbott
has drawn attention to governments' rash
decision to prioritise safety over prosperity.
"Whole societies have been locked down
essentially to protect the elderly."
Abbot accused politicians of not weighing
the cost of keeping the economy functioning
against keeping the elderly alive.
"Every life is precious and every death sad.
But that's never stopped families
sometimes electing to make elderly relatives
as comfortable as possible while
nature takes its course."
Calculating the impact on an economy
Abbot estimated coronavirus measures
aimed at keeping 1 elderly person alive
would cost $200,000 a year—which is
"substantially beyond what governments
will usually pay for life-saving drugs."

According to a May 2020 survey
many people worldwide believe Covid-19
has killed far fewer people than reported—
was deliberately created in a lab—
has symptoms spread by mobile 5G technology—
or is a hoax so doesn't actually exist.
Part of the YouGov-Cambridge Globalism Project
which questioned 26,000 people in 25 countries
the survey—designed with The Guardian—
found widespread scepticism that by May 2020
the virus had killed almost 1.1 million worldwide.
Nearly 60% of respondents in Nigeria agree

the total was "deliberately and greatly exaggerated".
An average 35% in South Africa Greece
Poland Mexico Hungary America
Italy and Germany feel the same.
That the virus was created and deliberately
spread by the Chinese government is believed
by over 35% of Turks Poles South Africans
Spaniards Brazilians and Americans.
Alternatively an average 25% of respondents
in Turkey Poland Greece France and Spain
believe the US government is responsible.
17% of US citizens agree.
Over 20% also think human-caused global
heating is a hoax the harmful effects
of Covid vaccines are being kept secret
and the 1960s moon landings were faked.

> "I know from my work in epidemics
> in Africa that where there is fear
> and panic and patients become
> isolated from their families
> it doesn't take long for rumours
> and fake news stories to circulate.
> In the Covid pandemic smartphones
> and social media have connected families
> who are separated because of the risk
> of infection—but they've also helped
> generate a blizzard of dangerous fake news."
> Professor John Wright cited one post
> that stated those with a health condition
> shouldn't go to hospital because
> "you will not come back alive".

A mass demonstration in central Berlin
drew an estimated 20,000 protesters
underscoring the allure of Germany's
burgeoning coronavirus-denial movement.
Marchers included neo-Nazis
waving the Reichsflagge
(adopted to replace banned swastikas)
dreadlocked peace protesters
and LGBTQ activists carrying rainbow banners.
They accuse the government and press
of lying about the pandemic.
Hardly any protesters wore masks
or followed social distancing rules
prompting authorities to put
an early end to the demonstration.
A clear majority of Germans support
Covid-19 restrictions and endorse
sanctions for those who break the rules.

A group of protesters unhappy with
the New Zealand Government's alert levels
gathered in Rotorua today.
One of the organisers explained
"It's about our freedom to choose
and whether we want vaccinations
5G or to be in lockdown.
It's only the flu season
so there's no need for lockdowns or masks.
People are actually killing themselves
wearing masks because they're
breathing in their own toxins.
This government wants us to be sick."

As of 22 May 2020 there have been
49 verified deaths of NHS staff from Covid.
The Guardian records 200 deaths
have actually been reported in the news
but it is clear many more have died.
Rebecca Mack (29) worked as a children's nurse
before going on to a job with NHS 111.
She had no known health problems
and fell ill after a work training session in Derby.
She was self-isolating alone at her home
when her symptoms worsened.
Mack called for an ambulance
and left the door open for paramedics.
They found her dead in her residence.

> It's not the accessory we asked for
> but it's the accessory we got.
> All over the world people are getting
> creative with their face masks
> which have become essential tools
> in preventing the spread of coronavirus.
> Now as warmer weather hits and beaches
> begin to reopen a new fashion item
> is making a splash: the trikini.
> Developed by Italy's Elexia Beachwear
> it's a bikini with a matching face mask.
> Hey if we're going to have weird tan lines
> we might as well look cute
> and be safe while we get them!

A Florida man and his three adult sons
have been indicted by a Federal Grand Jury

on charges of selling a toxic industrial bleach
they called Miracle Mineral Solution
as a cure for Covid-19 cancer herpes AIDS
diabetes Alzheimers and other conditions.
Prosecutors claim they established
Genesis II Church of Health and Healing
via which they received more than
$1 million in "donations" for the product.

As 50 million filed for unemployment insurance
American billionaires are $637 billion richer.
Mark Zuckerberg's wealth ballooned 59%
and Amazon's Jeff Bezos's 39%.
Since the pandemic began Big Pharma
has raised prices on over 250 prescription drugs
61 of which are being used to treat Covid-19.
Moderna has never brought a vaccine to market
but company insiders have sold $248 million
of shares—most after the company was
selected to receive White House funding.
Moderna will sell its vaccine for profit although
taxpayers footed its research and development.
In the most difficult economic crisis since
the Great Depression stock prices
are almost back to pre-pandemic levels.

Covid-19 has tipped 40 million living
in sub-Saharan Africa into poverty.
"These are countries with few social safety nets
high levels of debt and scarce resources
to respond to a crisis of this magnitude"
says Murithi Murtiga a Nairobi-based

analyst with the International Crisis Group.
"Household budgets have collapsed
due to the limited support for those who
lost livelihoods during lockdown."
The slow roll-out of vaccines in Africa
threatens to cause further misery.
The bulk of Africa's vaccines are supplied
via Covax and are manufactured in India.
But India has suspended vaccine exports
to cope with domestic demand.

> "I don't think I'll be jabbed.
> I don't trust them. Not even a little.
> I'm not anti-vaccine but
> this is moving too fast.
> What if it stuffs with fertility?
> I refuse even a Covid test.
> The symptom of death bothers me
> but do they care if you or your kid dies?
> As for those arrogant people who
> go out when they're sick
> in my view they need locking up.
> No thank you Dr Bloomfield.
> But guess what? I'm up for
> the challenge. Bring it on!"

Emergency room nurse Jodi Doering
declares "It's like a horror movie."
Her anguish is because patients who
are severely ill refuse to acknowledge
they're infected by Covid-19.
"I think the hardest thing to watch

is people looking for a magic answer.
Their last dying words are
'This can't be happening. It's not real.'
People want it to be influenza or pneumonia.
I've even had people hope it's lung cancer.
It just makes you mad and depressed
because you have to go back next day
and deal with it all over again."

An African doctor who advised citizens
against getting vaccinated has himself
been diagnosed with Covid-19.
Dr. Stephen Karanja served as chairman
of the Kenya Catholic Doctors Association.
He diverged from pro-vaccine bishops
by claiming mask-wearing and mass testing
could effectively stamp out the pandemic.
In a letter addressed to his fellow doctors
he declared "the vaccine for Covid-19
is unnecessary and should not be given."
He instead endorsed ivermectin
designed to treat parasites in animals.
No African medical association has
endorsed its "off-label" use of treating Covid
but demand has seen its price increase 15-fold.
Dr. Karanja was in ICU for seven days
before he succumbed to the coronavirus.

In June 2021 India's 31 million Covid cases
are the world's 2nd highest—behind the US.
Despite a nationwide vaccination drive
shortages of vaccines and logistical hurdles

mean only 8% of adults are fully vaccinated.
India's official death tally is now over 400,000
many dying during a brutal second wave
when people expired outside hospitals while
waiting for beds and bodies washed up
on the banks of the holy Ganges river.
Virendra Kumar had to bury his son
on the Ganges' banks because doubled
firewood prices had sent funeral costs surging.
With a family income under $100 a month
Kumar couldn't afford his son's cremation.

Despite Latin America having only
8% of the world's population
in April 2021 it has 35%
of the world's Covid deaths.
"Covid in Latin America is a story
that is just beginning to be told"
says economist Alejandro Gaviria
Columbia's former health minister.
"I have tried to be optimistic.
I want to think the worst is over.
But I believe that is counter-evident.
Patients arriving at hospitals
are now far younger and sicker
than we have previously seen."
A Peruvian man was fired
after contracting Covid.
Soon after his pregnant wife
gave birth to twins who shortly died.
"I left the hospital with one daughter
in a black plastic bag got in a taxi

and went to the cemetery.
There was no mass. No wake.
No flowers. Nothing."

As Brazil's Covid numbers rise steeply
President Jair Bolsanaro continues
to disparage coronavirus science
attacks the use of masks
and ignores his own health officials'
calls for a national lockdown.
In response to his country's soaring death toll
the President told Brazilians this week:
"There's no use crying over spilled milk."

With vaccination ongoing in wealthy countries
(which bought up the bulk of initial deliveries)
US companies Pfizer and Moderna
have raised their prices after clinical trials
showed their formula was more effective than
those of AstraZeneca and Johnson & Johnson.
Per shot prices range from S2.15 for AstraZeneca's
EU deliveries to Novavax's $20 deal with Denmark
to Moderna's price of up to $37 outside the US.
The 2021 first-half profit of BioNTech
—which co-produces the Pfizer vaccine—
has jumped to €4bn from €142m a year earlier.
Moderna generated a $4bn profit
on $6bn sales—the company's first half-year
profit since it was founded in 2010.

A mid-2021 survey found 23% of New Zealanders
say they're unsure or won't get vaccinated at all.

The Ministry of Health says that number
is lower than a few months previously.
Many affluent and well-educated professionals
are among those refusing to get a jab.
Former Prime Minister John Key recently
played golf in a foursome and two declared
they wouldn't be getting vaccinated.
"I was stunned" he said. "These people are
widely travelled and seem to understand
the risks of catching Covid—but still say no."
Sociologist Paul Spoonley observes anti-vax
rhetoric has gone full-blown conspiracy.
When holidaying recently he picked up
a local church newsletter and found an editorial
that undermined all current health advice
and ended by suggesting Covid involves
"an international plan to take over our country".

>"My name is Nemonte Nenquimo.
>The Amazon rainforest is my home.
>I am writing you this letter because
>the fires are raging still—because
>corporations are spilling oil in our rivers
>—because the miners are stealing gold
>and leaving behind open pits and toxins.
>Because the land grabbers are cutting
>primary forest so cattle can graze
>and the white man can eat.
>Because our elders are dying from Covid
>while you divide our lands to stimulate
>an economy that has never benefited us.
>Because we are fighting to protect

> our way of life—our rivers—
> the animals—our forests—life on Earth.
> It's time you listened to us."

International institutions have set the goal
of eradicating extreme poverty by 2030.
Despite "self congratulatory" messages
they are losing the fight says Philip Alston
the UN's outgoing special rapporteur
on extreme poverty and human rights.
"Rather than providing a roadmap for states
to tackle our era's critical problems
the energy surrounding the SDG* process
has gone into generating colourful posters
and bland reports that describe
the glass as $^1/_5$ full rather than $^4/_5$ empty.
Even before Covid-19 we squandered
a decade in the fight against poverty
with misplaced triumphalism blocking
the very reforms that could have prevented
the worst impacts of the pandemic."

[*Sustainable development goals]

"The Covid-19 pandemic has shown
how vulnerable the world is
to a truly global catastrophe.
But another bigger catastrophe
has been building for many decades.
In the past 10 years 83% of all disasters
triggered by natural hazards
were caused by extreme weather
and climate-related events

such as floods storms and heatwaves.
And they will only get worse without
immediate and determined action."
So states *Come Heat or High Water*
the 2020 World Disasters Report published
by the International Federation of Red Cross.
"The issues are not only financial.
It is time to shake off business as usual
and turn words into action.
We all—governments donors
the humanitarian and development
climate and environment communities—
need to act effectively before it's too late.
Let's not miss our chance."

In response to increasing climate disasters
the best the oil companies can offer
are vague pronouncements about
getting to "net zero by 2050"—
which is another way of saying
"We're not going to change
much of anything anytime soon."
The American giants like ExxonMobil
won't even do that.
French company Total has made
the 2050 pledge but is projected to
increase fossil-fuel production by 12%
between 2018 and 2030.
According to the Intergovernmental Panel
on Climate Change this is precisely when
we must halve emissions
to have any chance of meeting

the 1.5C temperature increase target
set by the 2016 Paris climate agreement
to which 196 nations signed up.

> "Every few years governments
> gather to make solemn promises
> about the action they will take
> to defend the living world.
> At today's virtual UN summit
> on biodiversity they will move
> themselves to tears with thoughts
> of the grand things they will do—
> then go home and sign another
> mining lease.
> At the last summit 10 years ago
> world leaders made a similar set of
> "inspirational" promises.
> Analysis published a fortnight ago
> showed that of the 20 pledges agreed
> in 2010 in Japan
> not one has been met.
> The collapse of wildlife populations
> and our life-support systems
> continues unabated.
> The world has now lost 68%
> of its wild vertebrates since 1970.
> It sounds brutal to say these meetings
> are a total waste of time.
> But this is a generous assessment.
> By creating a false impression
> of progress
> by assuaging fear and

fobbing us off
these summits are a means
not of accelerating action
but thwarting it."

"Build back better. Blah, blah, blah.
Green economy. Blah blah blah.
Net zero by 2050. Blah, blah, blah.
This is all we hear from our so-called leaders.
Words that sound great but so far
have not led to action."
So spoke Greta Thunberg during a speech
at the Youth4Climate summit held in Milan.
"We do need constructive dialogue.
But they've now had 30 years of
blah, blah, blah
and where has that led us?
We can still turn this around—
it is entirely possible.
But not if things go on like today.
We can no longer let the people in power
decide what hope is.
Hope is not passive.
Hope is not blah, blah, blah.
Hope is telling the truth.
Hope is taking action.
And hope always comes from the people.
Our leaders' intentional lack of action
is a betrayal of all
present and future generations."

Many young today feel the last century's
social contracts are broken.
"Most people my age are paddling so hard
just to stay still," says architect Tom.
"Nobody is asking for an easy ride
but many of my friends
are losing faith in the system."
Killian Mangan who graduated during
the pandemic last year and struggled to find
a job notes it feels as if "we are drowning
in insecurity with no help in sight."
A 20-something who works for
a central bank says "I sometimes have
this feeling that we are edging towards
a precipice or falling in it already."
A 30-something who works in private equity
says "The space I feel I occupy
in the sociopolitical order is akin to being
the first loss tranche in the debt stack.
Whenever anything bad happens
I have no doubt that because we lack
political and economic clout
we will be left holding the bag."
Akin Ogundele is the university graduate son
of immigrants and a born and bred Londoner.
Married with a family at 34 he feels trapped—
"My retirement plan is to die
in the climate wars."

> "When I think that it won't hurt
> too much
> I imagine the children I will not have.

Would they be more like me
or my partner?
Would they have inherited
my thatch of hair
our terrible eyesight?
Then I remember the numbers.
If my baby were born today
they would be 10 years old
when $1/4$ of the world's insects
could be gone
when 100 million children
are expected to be suffering
extreme food scarcity.
My child would be 23
when 99% of coral reefs are set
to experience severe bleaching.
They would be 30—my age now—
when 200 million climate refugees
will be roaming the world
when half of all species on Earth
are predicted to be extinct.
They would be 80 in 2100
when parts of Australia
Africa and the US
could be uninhabitable."

Sources

All excerpts have been cleared with the writers if required and where possible. I thank those who responded quickly and helpfullly. Excerpts from scientific research and media releases are used in accordance with International Creative Commons Licenses. I am especially grateful to The Guardian, which permits usage of short excerpts from their news articles, facilitating this work.

On 31 December 2019 Wuhan health officials ...
The data is drawn from *Timeline: WHO's Covid-19 response*, published on the World Health Organisation's website, and from Wikipedia's *Timeline of the Covid-19 pandemic*.

This is not nature's revenge ...
From *'We did it to ourselves': scientist says intrusion into nature led to pandemic*, written by Phoebe Weston, biodiversity reporter, published by The Guardian, 25 April 2020. Dr. Thomas Lovejoy is a Professor in the Department of Environmental Science and Policy, College of Science, George Mason University. Used with the author's permission and courtesy of Guardian News & Media Ltd. For more of Phoebe Weston's writing: www.theguardian.com.

Minister Chen Wen has clarified facts ...
From *Minister Chen Wen Gives Live Interview on BBC Radio 4*, a media release summarising Chen Wen's interview, published by the UK Embassy of the People's Republic of China, 26 April 2020, on www.mfa.gov.cn.

A series of monthly surveys were carried out ...
From *Animal sales from Wuhan wet markets immediately prior to the COVID-19 pandemic*, a study authored by Xiao Xiao, Chris

Newman, Christina D. Buesching, David W. Macdonald and Zhao-Min Zhou, published by Nature, 7 June 2021. For the full report go to https://rdcu.be/cIpI6.

The virus we're talking about ...
Edited from two speeches given by President Trump on 10 February, at the White House and during a rally in New Hampshire. Video and transcripts of his speeches appeared on multiple media outlets.

By mid-April 2020 1/3 of the world's population ...
The statistics are drawn from a summary collated by Global Health, published on www.thinkglobalhealth.org. Accounts of the Kentucky Covid party were picked up by news sources worldwide, along with Governor Andy Beshear's subsequent comments, which appeared in his Team Kentucky Covid updates posted to Facebook.

Former Australian PM Tony Abbott ...
From *Australia and the Coronavirus*, a speech given by former Australian Prime Minister Tony Abbot at the invitation of Policy Exchange. Responses to Abbot's speech were published by multiple media sites. This excerpt draws directly from the speech, a transcript of which is available at www.policyexchange.org.uk.

According to a May 2020 survey ...
This draws on a survey conducted by the YouGov-Cambridge Globalism Project, in conjunction with The Guardian. The raw survey results may be viewed on yougov.co.uk with the title: *Globalism2020 Guardian Conspiracy Theories*. Selected data was subsequently presented in a news article, *Survey uncovers widespread belief in 'dangerous' Covid conspiracy theories*, written by Jon Henley, Europe correspondent, and Niamh McIntyre, data journalist, published by The Guardian, 26 October 2020. Used courtesy of Guardian News & Media Ltd: www.theguardian.com. The total of 1.1 million deaths by May 2020 is from the John Hopkins' online

Coronavirus Resource Centre. By March 2022 total deaths are estimated to be at 6.1 million. However, some modelling suggests that, due to under-reporting, the true total may be 3 times that. See *COVID's true death toll: much higher than official records*, written by David Adam, published in Nature, 10 March 2022: www.nature.com.

I know from my work in epidemics ...
From an interview with Professor John Wright, medical doctor and epidemiologist, head of the Bradford Institute for Health Research, that appeared in *Bradford health expert condemns conspiracy theories over deaths*, written by Tim Quantrill, published by Angus and Telegraph, 19 April 2020: www.thetelegraphandargus.co.uk. Used with the permission of Professor Wright, whose institute's research may be viewed at www.bradfordresearch.nhs.uk.

A mass demonstration in central Berlin ...
From *German coronavirus deniers test Merkel government*, written by Matthew Karnitschnig, Chief Europe Correspondent for Politico, reporting from Berlin, published by Politico, 4 August 2020.

A group of protesters unhappy with ...
From *Covid 19 coronavirus: Lockdown protesters march in Rotorua*, written by David Beck, multimedia journalist, published by The New Zealand Herald, 5 September 2020. Used with the permission of the author. For more of David Beck's articles: www.nzherald.co.nz.

As of 16 April 2020 there have been ...
From *Doctors, nurses, porters, volunteers: the UK health workers who have died from Covid-19. A project to remember those who lost their lives working in hospitals, surgeries and care homes during the coronavirus outbreak*, compiled and written by Sarah Marsh, news reporter, published by The Guardian, 22 May 2020. The project remains available to view on www.theguardian.com. Used courtesy of Guardian News & Media Ltd.

It's not the accessory we asked for ...
From *Meet The Trikini, A Bikini That Comes With A Matching Face Mask*, written by Jamie Feldman, fashion and lifestyle editor, published by Huffpost, 18 May 2020: www.huffpost.com.

A Florida man and his three adult sons have ...
From a media release, *Florida Family Indicted for Selling Toxic Bleach as Fake "Miracle" Cure for Covid-19 and Other Serious Diseases*, published by the US Attorney's Office, Southern District of Florida, 23 April, 2021. This story was reproduced by multiple media outlets. The excerpt draws on the media release.

As 50 million filed for unemployment insurance ...
From *Robert Reich on how billionaires are profiteering off the pandemic*, published on Reich's blog and by Salon, 24 August 2020. Robert Reich is Chancellor's Professor of Public Policy at the University of California at Berkeley and Senior Fellow at the Blum Center for Developing Economies. He served as Secretary of Labor in the Clinton administration. His blog may be viewed at robertreich.org.

Covid-19 has tipped 40 million living ...
From *'We're just trying to survive': what Africa risks from a new Covid wave*, written by Jason Burke based in Johannesburg, published by The Guardian, 1 May 2021. Used courtesy of Guardian News & Media Ltd: www.theguardian.com.

I don't think I'll be jabbed ...
From an anonymous comment to an article on vaccination published in The New Zealand Herald, April 2021.

Emergency room nurse Jodi Doering ...
Numerous news sites picked up on Jodi Doering's views after a tweet she wrote went viral, and she was subsequently interviewed on CNN. The interview, from which this excerpt is drawn, may be viewed on www.edition.cnn.com, uploaded 16 November 2020.

An African doctor who advised citizens ...
Dr Karanja's views and his death were covered by multiple news outlets. Dr Karanja's letter is posted on Twitter. The material on ivermectin is from *Potential to save lives: An intensive care doctor argues for 'compassionate use' of ivermectin for Covid-19*, an op-ed written by Nathi Mdladla, published by Daily Maverick, 15 January 2021.

In June 2021 India's 31 million Covid cases ...
From *Poverty, stigma behind bodies floating in India's Ganges River*, written by Saurabh Sharma reporting from Lucknow, published by Al Jazeera, 2 June 2021: www.aljazeera.com.

Despite Latin America having only ...
From *After a Year of Loss, South America Suffers Worst Loss of Life*, researched and written by Isayen Herrera in Caracas, Sofía Villamil in Bogotá and Daniel Politi in Buenos Aires, published by WorldNewzInfo, 30 April 2021: www. worldnewzinfo.com.

As Brazil's Covid numbers rise steeply ...
From *'Out of control': Brazil's COVID surge sparks regional fears*, written by Charlotte Peet, freelance journalist based, published by Al Jazeera, 10 April 2021: www.aljazeera.com.

With vaccination underway in wealthy countries ...
From *Covid-19 vaccines: the contracts, prices and profits*, written by Julia Kollewe, published by The Guardian, 11 August 2021. Courtesy of Guardian News & Media Ltd: www.theguardian.com.

A May 2121 survey found 25% of New Zealanders ...
From *Well-educated and well-off anti-vaxxers are among us*, written by Virginia Fallon, published by Stuff, 23 May 2021. Used courtesy of Stuff Ltd: www.stuff.co.nz.

My name is Nemonte Nenquimo. ...
From *This is my message to the western world—your civilisation is killing life on Earth*, by Nemonte Nenquimo, published by The Guardian, 12 October 2020. Nemonte Nenquimo is cofounder of Eucador's Indigenous-led nonprofit organisation Ceibo Alliance, is the first female president of the Waorani organisation of Pastaza province, and was named one of Time Magazine's 100 Most Influential People of 2020. Her TEDtalk, *The forest is our teacher. It's time to respect it*, is available on www.ted.com. Used courtesy of Guardian News & Media Ltd: www.theguardian.com.

International institutions have set the goal ...
From *'We squandered a decade': world losing fight against poverty, says UN academic*, written by Peter Beaumont, a senior reporter on The Guardian's Global Development desk, published by The Guardian, 7 July 2020. Used courtesy of Guardian News & Media Ltd. For more by Peter Beaumont go to www.guardian.co.nz.

The Covid-19 pandemic has shown ...
From *Come Heat or High Water: Tackling the humanitarian impacts of the climate crisis together*, published by International Federation of Red Cross and Red Crescent Societies, Geneva. The full report is available at media.ifrc.org/ifrc/world-disaster-report-2020.

In response to increasing climate disasters ...
From *What Facebook and the Oil Industry Have in Common*, written by Bill McKibben, environmentalist, author and contributing writer to The New Yorker, published by The New Yorker, 1 July 2020. Bill McKibben is a founder of the grassroots climate campaign 350.org and the Schumann Distinguished Scholar in environmental studies at Middlebury College. Used by permission of the author. Bill McKibbon's website: billmckibben.com.

Every few years governments gather ...
From *Johnson's pledges on the environment are worthless. Worse is

how cynical they are, written by George Monbiot, published by The Guardian, 30 September 2020. Used by permission of the author and courtesy of Guardian News & Media Ltd. For more by George Monbiot go to: www.monbiot.com and www.theguardian.com.

Build back better. Blah, blah, blah ...
From an address given by Greta Thunberg on 21 September 2021 to the Youth4Climate summit in Milan, Italy. The summit was sponsored by the Italian government, who partnered the UK in running COP26. In her speech, Thunberg also stated: "They invite cherry-picked young people to meetings like this to pretend that they listen to us. But they clearly don't listen to us. Our emissions are still rising. The science doesn't lie." A video of her full speech is available on YouTube: www.youtube.com/watch?v=wpo33oLne-Y. Viewers' comments on her speech graphically illustrate the extent of paralysing collective stupification.

Many young today feel the last century's ...
From '*We are drowning in insecurity*': *Young people and life after the pandemic*, written by Sarah Connor, employment columnist and associate editor, Financial Times, published 25 April 2021. The interviews were gathered during an informal survey carried out by the Financial Times, which ran a global survey for under 35s asking them about their lives and expectations in the wake of the pandemic. 1,700 young people responded in the space of a week from countries as varied as South Africa, Cambodia, Norway, Australia, Denmark, the US, Portugal, Lebanon, Brazil, Malaysia, India and China. The full article may be viewed at www.ft.com.

When I think that it won't hurt too much ...
From *Why a generation is choosing to be child-free*, written by Sian Cain, Guardian journalist, published by The Guardian, 25 July 2020. Used by permission of Sian Cain and courtesy of Guardian News & Media Ltd. For more articles by Sian Cain: www.theguardian.com.

Boogying with the virus

A modest proposal

The animal park in Neumünster, 50 km north of Hamburg, has said it may have to begin feeding some animals to others to survive. Zoo director Verena Kaspari told Die Welt reporters that this "modest proposal" would have to be brought into action by the middle of May if the park continued to be shut. —Die Velt, news report

You know the situation's bad
when zoos start making plans
to feed some animals to the others
because Covid has cut supply lines.
Of course it's the right thing to do.
We need to ensure the future
economy of the zoo comes out
of this tough time
in a secure market position.
But which animals do we choose?
What about feeding the smallest
—those that don't really count—
to the biggest and hungriest
such as the cute furry bears?
Let's be realistic.
How many small scaled lizards
dating back to the Cenozoic
do we really need to look at?
The fact is when you go to the zoo
you can't see them anyway

because they're hiding under
a log or in a clump of grass.
Lizards have always taken social
distancing to a ridiculous extreme.
Who's really going to miss that?
And anyway they're survivors.
Lizards will be with us always.
The bigger problem is let a bear breakfast
on a few handfuls of squirming lizards
and after he's swallowed the last
he's looking round for lunch.
Then to feed the grumbling bear
we'll have to open bird cages and that's
tricky because people hate to picture
birds' delicate wings being crunched.
How about we feed a bear to the lizards?
Imagine how many lizards could survive
feasting on one rotund carcass!
Except—sorry I forgot—we can't do that
because bears are cute and marketable
but no one visits the zoo just to see lizards.
Or we could feed the bears to animals
that sell entry tickets themselves.
There's hyenas mongooses and eagles.
Fish seals and walruses.
Big cats are a massive draw
and always need a feed.
That way you could save so many
and all for the sacrifice of a few bears.
But we come back to the central problem:
take out the cute bears and
the zoo's attractiveness takes a nose dive:

it's a bear market out there.
How about the noisiest?
Those that hoot and screech across
the walkways warning others of danger
or just to announce "I'm here"?
They're so loud you can't hear the zoo's
director detailing future plans.
Looked at that way the zoo's noisiest
are a liability.
And personally I've never liked those who
screech over the top of everyone else.
Let's start sharpening the blades!
But I don't want to be pushy.
I appreciate a zoo reflects
our collective desires and not
everyone considers lizards expendable
just because so many of them
don't contribute to the zoo's bottom line.
So maybe it's too soon to decide.
Some choices are just too tough to make
when we're all profoundly stressed.
How about instead we pack the board
head for the nearest beach
and relax by going for a surf?
We can catch the waves rippling out
from Antarctica and across the globe
as its long-past-it Mesozoic glaciers
melt into the ocean.

We need to stop dithering

"Humanity is waging war on nature. Biodiversity is collapsing. One million species are at risk of extinction. Ecosystems are disappearing before our eyes."
— UN Secretary-General António Guterres

Okay. I get people think
humanity's impact on the planet
isn't wholly positive.
But there's another way to consider this.
Because when you look
at the situation with unbiased eyes
it's clear the planet is crammed
with way too many species.
We human beings need space
to spread out and do our thing.
But all these other creatures are
constantly encroaching on us
in the ocean on land in the air
hunting eating fornicating pooing.
And what do we get in return?
Life threatening bites stings rashes
and viruses so bad we have to shut
down the planet not to die.
It's out of whack.
I have a solution.
Let's evaluate all species and decide
which are really working for us.
If we can't milk it

enjoy looking at it
kill it for fun
or spread it on crackers
what the hell use is it?
Something has to give.
And while we're at it
I say we stop giving oxygen
to tree and dolphin huggers
who get goo-gaa about nature
when nature's set up for everything
to eat each other.
Wake up and smell the blood guys!
Nature's not a place to play footsie
with something that either wants
to tear you limb from limb
or burrow under your skin.
It's great you're loving nature.
But nature's not loving you back.

Of course having everything suffer
indiscriminately
can't be the game plan.
We need a strategy that's sensitive
to species' changing situations.
No one wants dumb creatures
to die in pain
due to our non-negotiable demand
we take over their habitat.
So how about this for an idea?
As we keep spreading across the planet
and our activities flush species
out of forests mountains rivers oceans

leaving them nowhere to live
I propose we capture them
then extract their genetic material
and store it in deep freeze.
Because we need to consider
our distant descendants.
Saving DNA gives them the option
of reanimating species they like
and putting them in
a post-Anthropocene theme park
where they can watch
big things eat each other and get
goo-gaa over what doesn't make it.
What could go wrong with that?
As for the specimens we capture
after we sample their DNA
I propose we kill and eat them.
That way they are protected from
a long distressing extinction.
And we get to dip something tasty
in batter and deep fry it.
What we can't eat we use to make
handbags shoes and medicine
for those who find their body
disappointing.
It's a win all round.

The big thing is we need to make
a decision soon.
Today. Tomorrow. The day after.
There's no time for typical human
dithering

where we argue endlessly
pretend to weigh the pros and cons
and in the end do nothing.
Displaced species are already on
their way out.
How many more times do we want
to watch a starving polar bear's
futile attempts to find food?
Let's show some compassion.
Let's make the tough decision.
But the right decision.
Let's put the planet out of its misery.

Hooray for the workers

The Walton family [who own Walmart] benefited enormously during the Covid-19 pandemic, making more than US$1 billion every week in 2020. A Walmart worker's starting wage is $US11 per hour.—News report

The RAND Corporation did a study of what they call the "transfer of wealth" from the working class and the middle class since the neoliberal assault began around 1980. Their estimate for how much wealth has moved from the lower 90 percent of the income scale to the very top is $47 trillion.
—Noam Chomsky

Rodney: "The peasants are revolting." King Fink: "You can say that again." —The Wizard of Id

Workers of the world
during lockdown you were
reduced to watching your days
through the window
as they loitered listlessly in the street
frustratedly hovering on the verge
of doing something.
Eventually they dribbled away
into the next month.
Or was it the next year?
(It's all a blur now.)
But whether pre- intra- or post-Covid

what difference has the pandemic
really made to your life?
(Assuming
because you're reading this
it hasn't killed you.)
Remember the lockdown workers
—the cleaners and security guards—
in Melbourne's isolation hotels
who needed second jobs because their
essential work didn't pay a living wage?
They shed virus taking public transport
to their next workplace.
How about the applause for essential
British hospital workers who laboured
in virulent wards with insufficient PPE
and died in droves?
Or the Italian opera stars who sang
from their balconies
their talents spreading joy though
streets and the mediaverse
in inverse proportion to their
prospects for earning a living?
During lockdown's immediate aftermath
while office-based employees were
encouraged to enjoy working from home
and the low-end of the professional class
continued their previous grind
of being overworked and underpaid
our nation's working underbelly
(if they had work at all)
remained locked into the same
low-paid jobs or zero hour contracts

or couldn't escape the gig economy—
the keys of their self-supplied cars
controlled by unknown overlords
living in distant countries
stashing their profits in tax havens.

Workers of the world it's time
we all faced up to reality.
The brutal truth is the class war
is over.
The workers lost.
And no government sufficiently cares
or possesses the will (or courage)
to take a deep dive into
the economic and ideological drivers
that pilot today's escalating inequality
and have side-swiped your family's future.
Perhaps this diatribe should pull up here
back into a cul-de-sac and turn itself off.
Maybe we should appreciate
what's left of our days
accept the purpose of modern economics
is to enrich the wealthy
and happily ride our future into whatever.
But here's the thing.
It's no accident.
This situation was engineered
40 years ago ...
when Treasury advised a high rate of
unemployment was an economic good
that reduced labour costs and grew profits ...
when unions were targeted to erode workers'

protections so they could be excluded
from sharing in those profits ...
when political parties jostled to brand
the unemployed bone idle to justify
Ruthanasia and to keep a tight lid on
the costs of governmental social support
(knowing their policies would bake in
high levels of sustained unemployment) ...
when migrant workers were brought in
and paid minimum to curb wage growth
then were demonised and raided when
politicians went electioneering ...
when left-leaning intellectuals defended
the battered workers by devoting decades
to vehemently arguing over who had
the most authentic identity.
In the tumult the business sector
went in and hoovered up all the cash.
Today Big Ag Big Pharma Big Oil
and Big Utilities—backed by border-
crossing equity funds and banks—
declare record profits
while workers flap helplessly
in unyielding economic headwinds.

Workers of the world
no one is even trying to pretend
any more.
The plan all along was to reduce you
to a state of serfdom
so the uber-rich could grab more of what
they already had in plenty.

Sure 21st century serfs own mobile phones
computers SUVs and gigantic 4K LED TVs.
They've also been maneuvered to borrow
from banks to overpay for an education
which makes them the smartest (indebted
and so indentured) serfs in history.
Occasionally an exceptional serf
is even made a knight or dame—
just to prove an egalitarian pulse beats
in the sphincter of the modern world.
Just don't get too hopeful your kids will reach
the bottom rung of the housing ladder.

Yet workers of the workers don't fret!
There's good news for you
in the looming post-Covid world.
The moneyed will always need serfs.
You are an essential cog in
the machinery of their daily life.
Consider climate change not a catastrophe
but an opportunity offering exciting
new work opportunities ...
manning barricades to keep out
climate-displaced migrating millions ...
piloting water taxis between the 10%'s
gated and guarded communities ...
mining lithium in ocean sanctuaries
for batteries to run amphibious Teslas ...
transporting effluent by hand because
the sea is inundating sewage systems ...

Workers of the world be relieved
you've luckily survived Covid.
You just need your luck to hold out
a little longer
so you'll be among the favoured serfs
the wealthy select to ensure
they comfortably ride out the turbulence
of these late Anthropocene days.
Workers of the world reject
the mindset you're staring down
the barrel
of an inhospitable future.
Be content your future hinges on
correctly answering just one question:
"Do I feel lucky?"
Well do you—
serf?

Psalm for the End Times

"The world is ending. That is why the climate is changing. ... I can try my best to change it but I can't. Everything happens according to God's will. ... We should keep going to church and worshipping God. If we do that we will go to heaven."
—Interviews with Papua New Guineans

"I believe we need Christian men and women in [political] office today. Now more than ever in the history of the church, we see the End Times."
—Mississippi Secretary of State Michael Watson

I begin with a confession.
I don't expect the Big Guy will place
my name tag on the heavenly table
He's setting for those who have
done whatever to join whoever
that He'll assign to sit wherever—
I have too many howevers.
Nonetheless at this juncture in history
when a virulent foe is marching through
the world heartlessly declaring
I'll have that one!
And that one!
That one too!
what happens after
the pitiless touch takes us
is surely worth considering.
We all want to believe those we love

will end up somewhere good
(or at least not bad)
when the maw of death opens
and sucks them into
What
(inevitably for us all)
Comes Next.
In the meantime I've got a question.
What about *before* we die?
Isn't that even more important?
Right now?
Faced as we are by a biological burp
smack in our faces.
How about we give that bell a ring?

So what *is* going on?
Picture humanity crammed into
a gigantic kick-ass double cab ute
burning through fossil fuel
like there's no tomorrow
our symbol of what's important
—crucifix tractor Buddha furry kiwi
bullet dollar sign fluffy red dice—
bouncing under the rear-view mirror
as we drive like crazy
veering right then left
indifferent to the roadkill
wiping out native frogs on one corner
endangered parrots on the straight:
SMACK! coloured feathers
floating in the dust behind us.
But it's a cantankerous ride

because we can't stop arguing over
what exactly—out the windows—
we're seeing.
And where the hell we're headed.
But that was always the Big Guy's plan.
Right?
He left us to our own devices way back
when He tore up
Adam and Eve's tenancy agreement
and turfed them out
onto the Mesopotamian plains.
Of course He had a few bottom lines.
Don't steal don't lie don't kill
dedicate a day a week to being holy
and if you want to swear don't involve Him.
Oh and don't shag the neighbour's spouse.
All very straight forward.
But He wasn't telling us what to do.
And once Cain modelled humanity's
future behaviour by murdering his brother
there was no stopping us.
Just look at what we've done to the place
since!
But the omniscient Big Guy knew all along
we would reach today's apocalyptic phase
of our pedal-to-the-metal journey.
Right?

So how do we respond to
the pressing problem of climate change?
Besides buying kayaks and fire extinguishers?
For issues this serious it's best to call on

those in the know.
That's the religious.
They haven't just spent 20 centuries
expecting the end of the world
(any day now)
they're striving to get elected so
they can ensure the apocalypse
has a dedicated budget sufficient to
deliver fitting explosive pizzazz and
(what's become an embarrassment)
within a credible time frame.
They're helped by the theatre of
the End Times having sped up.
It's now playing 24/7 in a flood of fires
and famines fanned by decimation
exploitation repression and wars—
all gingered by a newly minted pandemic.
Who needs 4 frenzied horsemen
heralding the apocalypse?
Trust humanity to get the job done.
What do we want? The End Times!
When do we want it? Now!

Yet—and in the face of these certainties—
I feel obliged to raise one small doubt.
Some among us empathise with our cousins
whose low-lying Pacific islands are battered
by increasingly massive cyclones
their houses knocked over by storms
salt from rising sea level
withering their crops' roots.
Since the islanders have little

and their future prospects offer less
what can they realistically do
but hope for better in the afterlife?
If I was them I'd take that deal.
Multinational corporations aren't
bringing anything to the table.
Nor are wealthy populations (like ours)
who talk fancy but are too addicted
to a carbon-fuelled lifestyle
to change their ways anytime soon.
Complicating the picture is that
the Big Guy has been ghosting us
ever since our forebears showed
what they thought of Him
by nailing His Son to two hunks of wood.
His subsequent modus operandi has been
to act unseen through earthly agents
a motley crew of prophets priests
and the inspired who look good on TV—
which makes oil companies divine proxies
terminating the planet on His behalf.
But here's where my small doubt lands.
Is wiping us out still the plan?

A literature review reveals
the Big Guy's last publication
was almost 2 millennia ago.
It's a revelatory text
declaring after universal conflagration
the celestial family will take up
official residence on Earth.
(At least what's left of it).

It's a poignant promise.
But what's happened since?
The delivery timetable has blown out
so completely
that on humanity's behalf I have to ask—
is annihilation still the objective?
Or has the Big Guy changed His mind?
Is He even still paying attention?
Could His lack of published updates
signal He's relinquished His tenure?
The answer is not academic.
Right now we're hurtling
full throttle down the highway
with calving glaciers and forest fires
looming dauntingly into view.
But if He's lost His enthusiasm
for smoking the planet
then our efforts to go hell for leather
no longer have His sign-off.
So have we been abandoned?
Are we crashing the planet on our own?
Since the Big Guy's current answer is
superterrestrial silence
maybe we should ease our foot off the pedal
and give ourselves a little time out
to consider
What Comes Next.
And I don't mean in the distant future.
I mean right here right now!

For those serious about our future
the gloomy view is humanity

is impotently doing donut burnouts.
Yet why be so pessimistic?
I'm with the religious:
life is more than just something to do
to fill in the time
between birth and death.
What we do matters.
But without a clear recommitment
from the Big Guy that burning it all down
then flooding what's left
is the way to go
I feel obliged to unequivocally state
—Right here! Right now!—
that as a collective planetary goal
going all out for the End Times
doesn't grab me.
I'd prefer another option.
Like survival.
Giving the generations who follow us
a chance
just seems the fair thing to do.

But what do I know?
Maybe the Big Guy has a secret plan
so amazing—so extraordinary—
but so inscrutably complex—
it's even got Him flummoxed how
to present it in plain writing.
Well Big Guy I have good news for You.
Humanity has reached
a climacteric crossroads.

If You have a new plan for
What Comes Next
now's the time to let us in on it.
Have faith that whatever it is
we'll get it done.
We're revved and ready.
Just don't keep us waiting
too long
(our tyres are almost burned out).
Hey Big Guy—can You see us
down here
honking our horns?
We're expectant. Eager. And waiting.
Your move!

Let's give it another go

In the 1950s about 75% of Americans trusted the federal government to "do what is right". The Pew Research Center's 2019 poll revealed trust in the US government was down to just 17%.
—Figures quoted in The Hartmann Report

80% of Australians and 83% of New Zealanders agree government is generally trustworthy, up from 49% and 53% respectively in 2009.
—Australian and NZ university research

Look. I know no one's in the mood.
Something we can't even see
has totally screwed with our lives
setting off huge waves of anxiety.
This moment in human history
is not one we ever imagined occurring
in our modern high-tech world.
We urgently need to turn things round.
So I have a suggestion.
Let's give democracy another go.
Okay. I admit it didn't turn out so well
last time
before the pandemic struck
all those wars market meltdowns
terrorists wanting to kill us
and governments trying to stop them
by spying on everyone.
But isn't being spied on the trade-off

for being free?
My point is it's not the only way
to do democracy.
I think we can make a better fist of it
second time round.

Of course democracy has its downsides.
Giving everyone a vote is great in theory
but in practice too many don't engage.
To start with the young.
They're way too big a demographic
and they keep asking awkward questions
no one can answer like
*You've been able to live your future
but what about my future?*
and *Why do I have to live
in the messed up world you made?*
Kids right? All talk no pragmatism.
But don't worry about them.
As they get older
and learn how the world works
those who survive droughts
storms fires landslides
and water sucking round their waists
will stop moaning and come round.

A bigger worry are the idealists
those who think they know what's best
for everyone else.
Do-gooders have no idea how much damage
they do to how things are.
Of course we respect their right

to march and shout themselves hoarse.
We're all better off for them doing that.
But they need to appreciate we can't
go cold turkey and jump into a future
that ditches longstanding and
proven processes outright.
Who knows how that world will look?
We created democracy to ensure
personal freedoms
and to provide stability.
It's a system of check and balances
designed to stop us crashing
our world.
Okay so that hasn't worked out.
But it's why we need to change.
Why we need Democracy Mark II.

Balancing pushy idealists we have workers.
They're pragmatists.
They have to be because their greatest
responsibility is their kids.
Workers want the world to be a ladder
their children can climb to a better life
not the slide they've been on
the last 40 years.
But workers don't have time to read
fence-sitting official reports
or digest 1000-page longitudinal studies
that provide the information they need
to ... what?
Acknowledge factors beyond their control
will dictate their kids' future lives?

They already know that.
They just want our leaders to lead us
out of this into
something—anything—better.

Which brings us to the elephant
squatting in the middle of the room.
Leaders.
Now more than ever we need
percipient visionary empathetic leaders
who see how life is and have a plan
to create an equitable future.
It's obvious where they need to start.
Reforming democracy.
As we would expect America
is showing inspirational leadership
in this space.
Sober-minded politicians are sensibly
clearing electoral confusion by ensuring
only those who back them
are allowed to vote.
Meantime China has shut Hong Kong
and silenced journalists and dissenters—
Russia is assassinating opponents
and cyber-attacking everyone else—
Africa is being stripped of its minerals
and using wars to draw media attention
so no one notices its leaders are siphoning
aid money into Swiss bank accounts—
the Haitian president is shot dead
in his home—
New York's subways are flooded—

Henan China is flooded—
Central Europe is flooded—
the Amazon is burning—
Siberia is burning—
North America's west is burning—
and billionaires are entertaining us
by racing each other to launch rockets
and give the elite space flights
so they can do what they love best ...
look down on the rest of us.

Yeah nah yeah yeah nah you're right.
What have I been smoking?
Democracy Mark II is dead on arrival.
It was founded by the ancient Greeks
so of course it's irrelevant now.
We've out-evolved it.
What we need is a political process
adapted to today's complex world
to replace decayed democracy.
We have oodles of options:
plutocracy technocracy adhocracy
corporatocracy cyberocracy kleptocracy—
this last better known as bancocracy.
There's wide support for theocracy
especially the upmarket TV evangelist
Old Testament version.
How about a military-led stratocracy?
Or ochlocracy geniocracy pathocracy?
(We're agreed no one wants noocracy.)
Past successes worth reconsidering
include feudalism elitism nepotism

communism totalitarianism fascism.
There's monarchy oligarchy anarchy.
Or throwing our hands in the air and
reverting to a banana republic.
Surely any is preferred to the car wreck
we have now.
The question is—how to proceed?
I know. Let's identify those we think
best fit today's world and create a list.
Then we could—dare I suggest it?—
organise a vote!
Yeah nah yeah yeah nah just kidding.
What we really need is a coup.
Put today's lot up against the wall
bulldoze the bodies
then invite pragmatic visionaries
—billionaires preachers ex-presidents—
to show us what real leadership looks like.
Give them the freedom to build
the world we all deserve.
Whew! Good to get *that* job done.
Solving the big problems is tough work.
What's next?
Anyone up for sexism? Ageism? Racism?
How about the optimal price for
cheese on the moon?

The world needs therapy

The individual's soliloquy
[After Nicanor Parra]

I'm the individual.
For a long time I floated in warm darkness
feeling nurtured and loved.
Then I was ejected into a world
of harsh light and hands that
beat me till I screamed.
I'm the individual.
People loomed in my face
jumped me on their knees
jabbed fingers into my cheeks.
I learned to grit my teeth.
I'm the individual.
By giggling and crying I found
I could get what I wanted.
I learned to manipulate the oohs and aahs
that flitted all day before my face.
I grew chubby on my new skill.
I'm the individual.
Next came language.
Through a natural process
of neurological and cultural osmosis
sound merged into the names
for people and things.
I especially liked the names for me.
I'm the individual.
Simultaneously I discovered legs.
I learned to walk towards what I wanted
and run from what I didn't.
I felt the consequence of being caught.

I'm the individual.
Soon after I discovered the difference
between what is mine and not-mine.
That just because I held something
didn't make it mine.
Or stop others taking it.
Or me taking it back.
I'm the individual.
Without being asked
I found myself being educated.
My tongue and mind were stuffed
into a pedagogical meat grinder
and mashed into predetermined shapes
to help me say and think what
everyone else said and thought.
I'm the individual.
Later I was dumped on the streets
and instructed to wrest a living from
the whirl of commerce.
If I did not willingly take part
I would be stigmatised a loser.
I didn't realise it then but
work is institutionally structured
to make other people rich.
I was forced to work not to avoid stigma
or to make others wealthy
but because I would go hungry and cold.
I'm the individual.
Okay.
After a while innate biological urges
mated with social imperatives
and one morning I woke up married.

Soon children shrieked in the yard.
The imperative now was to earn piles
of money money money.
Which never stayed around for long.
I'm the individual.
More years passed.
Marriage became a metal bucket
stuck in the middle of the hallway
which I kicked my toe on
as I left for work each morning
and again when I returned at night.
The tap that dripped into it
was the non-stop demand for
money money money.
I'm the individual.
Eventually cold water was poured
on marriage my family dissolved
and I was washed back into the streets.
But I couldn't be left alone.
A lawyer required my presence.
As I signed away the house he said
at least I still had decades of work
to look forward to.
He then handed me a bill demanding
yet more money money money.
I'm the individual.
For a time I enjoyed the convenience
of living in a boarding house.
As a break from the damp smell
I took up walking the streets at night.
Looking in the windows of
eye-wateringly expensive restaurants

I realised even a rock star economy
is constructed so only a few will ever
stockpile money money money.
The rest of us are on the outside
looking in.
I'm the individual.
That's when I decided I needed
to look though the keyhole.
I needed to do my own research.
I'm the individual.
I learned you have to ask questions.
And answers are never forthcoming.
Vested interests make sure of that.
Helped by the media.
But others can show you the way.
I'm the individual.
Okay then.
To cut an involved story short
the keyhole turned into a mouse hole
which became a slide
and I found myself spiralling
into a rabbit hole.
I found a big crowd down there.
All free-thinking individuals.
All upset about the way things are.
All offering me rabbit pie.
I'm the individual.
So I joined a troupe of rabbiteers.
We marched in a mass protest.
We camped in the liars' backyard.
We demanded every one resign.
I might have jostled a policeman. Or 3.

We refused to leave until we had
freedom to do what we want
when we want.
The media misrepresented everything.
I'm the individual.
Presently I received texts saying
my kids' school fees were due
and my rent payment had bounced.
Money money money!
I'm the individual.
I left the freedom of Rabbit Camp
taking vindication with me.
To show how right I was
I closed my eyes and drove through
the stop light at an intersection.
In the rear-view mirror I saw
my action had absolutely no bad effect.
That proved how much people live
in fear of what will never happen.
I'm the individual.
The ex-wife phoned to tell me
my kids loved and missed me.
I knew this was a coded demand
for money money money.
I'm the individual.
My children and I had fun.
Next day I felt sick and missed work.
Then my eldest fell sick.
And the youngest.
They must have contracted virus
from their schoolmates because
several of them fell sick too.

My youngest went to hospital.
The ex-wife claimed he almost died.
This was obviously fake news
concocted to hurt me.
I'm the individual.
From that day a huge metal bucket
sat in the ex-wife's driveway
keeping me from my kids.
I recovered and returned to work.
For a while I lost my smell and taste.
Including my taste for rabbit stew.
I'm the individual.
Alright.
Months passed. The world turned.
I look back and wonder what happened.
Perhaps I should have chosen other work.
A different wife.
Had different kids.
Maybe I need to get back to the warm
darkness and start over.
But I still wouldn't be free.

America

[After Allen Ginsberg]

In May 2020: The US climbs towards 100,000 Covid deaths; President Trump says people want to go back to work, while admitting that would lead to bodies "filling Yankee Stadium"; surveys reveal 7 in 10 Americans are against the President's proposal; Tropical Storm Arthur is the first of what will be a record 31 storms in the 2020 Atlantic hurricane season; detentions on the Mexican border double; police officer Derek Chauvin kills George Floyd by kneeling on his neck; Black Lives Matter street protests erupt in 450 cities around the country; the President tweets protesters are "thugs" and shooting them may be needed; that tweet is flagged by Twitter for promoting violence; in retaliation the President threatens to close all social media platforms; the US has 750 military bases in 80 countries; in 2019 it dropped 12,000 bombs and missiles (averaging 32 a day) in North Africa and the Middle East; a Pentagon report released 4 May records just 132 civilian deaths in 2019 due to US military operations in Iraq, Afghanistan, Somalia and Syria, and none in Libya or Yemen; the President inveighs against "criminal" Democrats, accusing them of treason; the US faces an invasion of Asian giant hornets.

America what the hell has happened to you?
America 10.35 p.m. May 29 2020.
You used to lead the free world.
Now we're all scratching our heads wondering what's put
    you in a huff.
Is it us? Is it something we said?

Are you just not that into us any more?
Or are you so into yourself you don't even realise we're
	trying to catch your eye?
America I want to know why you're in a ditch hitting
	the accelerator and making us wince at the sound of
	ungracious whining.
I'll wait for an answer because I see you're busy right now
	blaming others for driving you off the road.
America I only mention this because I'm trying to draw
	your attention to that burning smell.
Is that you? Are you on fire?
I ask because you're upside-down your wheels are spinning
	your engine is screaming and you're shrouded in
	smoke.
America there's no delicate way to put this.
Just where the hell do you think you're going?

America we need to talk things over.
I know you have other things on your mind.
But that's exactly what I'm concerned about.
We need to talk about your mind.
Because I'm no longer sure about your intentions.
You always had attitude but now you've put on a bully boy
	pose and are threatening everyone.
You don't seem to like anything any more.
Except money burgers and guns.
America you used to be the world's straight shooter.
Now you don't care who you shoot.
Why are you so obsessed with munitions?
No other country lets masked men armed with military
	weapons march into a town hall to proclaim their
	right to do what they want.

Your idea of freedom has made you world leader in mass
      shootings.
Perhaps you're nostalgic for the grand old days when you
      were a cowboy and shot everyone who didn't have a
      white skin?
I suggest this because your police keep killing black men
      and when your populace protests they are met in the
      streets by lines of strapping stormtroopers.
From the outside it all looks ominous.
You're not safer just because every asshole is armed.
Just like the world wouldn't be safer if every country had
      a nuclear bomb in its hip pocket.
It doesn't take a genius to know where this is headed.
The thing is the rest of us are trying to get on.
We have our differences.
And we don't always enjoy each others' company.
But we're not getting locked and loaded over it.
We're making an effort to work it out.
America why have you crossed everyone off your
      Christmas card list?
You're acting like you're the only country that counts.
America I'm seriously worried.
You have the most bombs by far on the planet.
And you've put psychos in charge.
Forgive me America but I have to speak bluntly.
We're all terrified of what you're going to do next.
America you have to sort your shit out!

America I love your music art poetry story-telling and
      life-enhancing technologies.
I love how you throw everything against the wall of
      culture then see what sticks.

America a hell of a lot has stuck with me.
Blues abstract expressionism Hollywood Miles Davis
    Walt Whitman Joan Didion Andy Warhol.
And you landed a freaking tin can on the Moon!
Then there's all those foreigners you made great—
Charlie Chaplin Audrey Hepburn Alfred Hitchcock
Helena Rubinstein Ayn Rand Madaleine Albright
Albert Einstein Nikola Tesla Werner von Braun.
There's so much to admire you for.
But don't get me wrong.
I'm not crazy about all you create.
Your obsession with celebrity and bling is childish.
Your rampant materialism sexism racism and militarism
    are ugly.
You have this weird religiosity that's so elastic no one
    blinks when politicians and TV preachers use it to
    support the next cock-eyed fear-filled slur.
Enabling corporations domiciled in foreign tax havens
    to asset strip the planet so your 1% can buy an
    Armageddon hide-out in New Zealand isn't cool.
And your primetime sitcoms leave me cold.
Humour derived from putting people in awkward
    situations to laugh at them suggests profound anxiety
    about your self worth.
Then there's your comic book concept of heroes.
It's time you realised your national narrative that wrongs
    are best righted through violence isn't helping anyone.
Including yourself. Especially yourself.
Let's face it—your obsessive creation of endless chases
    fights and shoot-outs is demented.
Once upon a time you told stories about good guys
    making the world safe from bad guys.

Now your stories are about one good guy surrounded by hordes of bad guys.
What are you saying?
That you're the lone good guy in a nasty at-you world?
America this may come as a shock but we're all human beings sharing this planet together.
Of course the big question is who is the bad guy you're so terrified of?
I've regretfully come to the conclusion it's you.

America when I was young everyone thought you were the best guy.
You were welcoming took your place in the world seriously and for a while you even engaged in detente.
Now you keep your distance and only open your mouth to tell the rest of us to hand over our money.
Meanwhile you're pretending your burning streets are caused by aberrant individuals and have nothing to do with you.
America even your own citizens think that's crazy.
They spend gazillions on drugs and therapy to deal with the stress of walking out their front door each day.
America do you think visiting a shrink would help you deal better with the stress of being you?
What I'm getting at is I run hot and cold on you.
But I'm willing to sit down and talk it over.
Are you?

America please don't go paranoid on me here.
Don't see me as a bad guy who has it in for you.
All I want is for you to get back to the best version of you.

Meanwhile well-meaning citizens are carrying on as if no skeletons are falling out of the closet.
Others are attempting to shove them back in while blaming the last guy who was in the room.
But that shtick has had its day.
You're kidding yourself if you think all this will magically melt away.
There are too many closets and way too many skeletons.
America have you got your phone out again?
Swiping left or right won't change a thing.
Nor will trying to make yourself feel better by digging another oil.
Because the pipes in the basement are bulging.
America we're all afraid that if you don't take this seriously the whole place will blow.
None of us smelling the smoke drifting across your borders will take any pleasure from that.
Believe me America when I say you're not the only one who has to take their temperature seriously.
The entire planet's basement pipes are bulging.
We all need to work together to turn the situation around.
Tell me America what are you going to do?
We're on the clock so don't take all day.
America I'm inviting you to rejoin the world and put your exceptional shoulder to the wheel.
America are you listening?
America can you hear me?
America are you still there?

Looky yonder

[After Led Belly]

Whoa! Black Betty.
Bam ba lam!
Betty had a baby.
Bam ba lam!
Wrapped it up in orange.
Bam ba lam!
Sent it off to preschool—
put it in a playground—
showed it other children—
never got along—
Bam bam bam ba lam!

Betty had a chancer.
Bam ba lam!
Showed it to its father.
Bam ba lam!
Learned to do business—
art of the tax dodge—
bankers pay for buildings—
suckers pay their bills—
Bam bam bam ba lam!

Whoa! Black Betty.
Bam ba lam!
Betty had a billionaire.
Bam ba lam!
Thought to run for President.
Bam ba lam!

Tapped people's anger—
paid off surly porn stars—
smeared loser rivals—
hung in Epstein's bar—
Bam bam bam ba lam!

The country had elections.
Bam ba lam!
They were the greatest ever.
Bam ba lam!
Lost to Crooked Hillary—
won the White House office—
loved the public licking—
the world went out drinking—
Bam bam bam ba lam!

Whoa! Black Betty.
Bam ba lam!
Betty's baby burst its nappy.
Bam ba lam!
Peed on all the people.
Bam ba lam!
Cheated on the golf course—
set its kids in office—
snubbed voters' losses—
splashed in the swamp—
Bam bam bam ba lam!

Betty had a President.
Bam ba lam!
Had to do it again.
Bam ba lam!

Downtalked mass infection—
favoured one complexion—
sought to steal the election—
encouraged insurrection—
Bam bam bam ba lam!

Whoa! Black Betty.
Bam ba lam!
Whoa! Black Betty.
Bam ba lam!
Baby had a meltdown.
Bam ba lam!
Plans the biggest rebound.
Bam ba lam!
Backed by the godly—
cynics and crazies—
fans a national war—
wants more more more!
Bam bam bam ba lam!

Looky looky yonder
looky yonder
a dark white cloud
has blocked the sun.
Looky looky yonder
looky yonder
an orange wolf
in a suit has come.

How to found a nation
[To the memory of Āpihai Te Kawau]

Accept a commission from the Crown
to explore the globe in its name.
Crew and provision a ship
then courageously sail the seven seas.
Appreciate the sole measure of success
is claiming new lands for the Crown.
When you find the best options
are already annexed by Belgium
France Netherlands Portugal Spain
refuse to be disheartened.
Continue sailing a southerly course
into the planet's most distant regions.
In the vast southern Pacific Ocean
discover two islands sufficiently large
they take days to sail around.
Permit yourself an exhale of relief.
Claim the islands by raising flags
in your Monarch's glorious name.
Note the presence of natives
chanting along the shore.
If you consider it appropriate
shoot a few of the rowdiest to show
your intentions are peaceful.
Sail away satisfied at your success.
Report on your endeavours anticipating
the glee of those who commissioned you.
Deservedly end your days being cooked
on a deliciously hot Hawaiian beach.

Rest easy knowing Crown officials
meeting in the gilded offices of Empire
are so enthused by your report
they appoint an out-of-work naval officer
as the first Governor tasked with hewing
that unmade land into a colony worthy
of a place in the Commonwealth Crown.
Be comforted that Governor will select
administrators bankers and lawyers
skilled at presenting to the native Māori
the dazzling advantages that will inevitably
follow their joining the modern world.
As American and French traders circle
observe the Governor successfully
persuade reluctant chiefs to sign
a bilateral treaty ceding sovereignty
to your glorious Monarch by arguing
they would otherwise lose their lands
to unprincipled and conniving foreigners
acting entirely in their own self-interest
who lack the Crown's fairness or
its scrupulous attention to detail.
That the Māori version of the Treaty
does not match the English version
view as regrettable but insufficient
reason to interrupt native absorption
into the expanding Empire.
Worry with the put-upon Governor
as he strives to fulfil his assigned task
of establishing a viable colonial outpost
in the face of the Empire's administrators'
refusal to allocate him any funds.

Applaud a far-sighted chief generously
sharing land sufficient to build a capital.
Feel the governor's joy as he establishes
that capital on a great harbour
far from the unconstrained settlers
carving up native lands down south.
Watch him borrow from bankers
then use "fair and equal" contracts
to purchase more Māori land
at a price he considers scrupulously fair
which he promptly on-sells to settlers
at an impressive profit
thereby raising the capital needed
to finance the extension of Empire.
Approve the synchronous actions of clerics
as they convert natives to Christianity
then persuade these new believers
to transfer vast tracts of land
to the caring control of the Church.
Thereafter feel his profound compassion
as thousands of Māori expire
from influenza smallpox measles syphilis
inadvertently shared by those newly arrived
from the sublime centres of Empire.
Feel his pain as his own health is decimated
by the settlers' manoeuverings
that vexatiously contest his tenuous office.
Lament his premature death and burial
but glow with gratitude acknowledging
the sacrifice he made to establish
Crown control of this rough and ready land.
Philosophically note the human cost

that inevitably accompanies
the gifting of modern advantages
to colonised and colonisers alike.

Sigh as the cycle begins again
with the new Governor forced to
grapple with the ongoing consequences
that attend new settlers' need for land.
Feel musket bullets plow flesh
as wary natives and bumptious settlers
disagree on who owns what.
Watch disagreements escalate into
skirmishes murders massacres.
Regret the over-zealous actions
of British Army officers leading
ragged bunches of colonial conscripts
who occasionally pursue the harsh path
of extermination rather than assimilation.
Feel the Governor's anxiety fall only when
Parliament passes a Native Land Act
which provides a process for land
to pass equitably from
colonised to colonisers' hands.
Watch approvingly as shrewd Māori
enter into leasehold agreements
with farming colonisers that advantage
both sides economically and savvily
allow the natives to use colonial troops
to halt incursions into their territory
by rapacious neighbouring tribes.
That those leaseholds are soon forgotten
and farmers use Court judgements

to claim the land outright
regard as a necessary lesson for the naive
regarding the machinery of Empire.
That dissenters who do not wish
to sell their land must argue their case
in courthouses located far from their homes
using an as yet unmastered language
view as necessarily reflecting the will
of those who now legally rule the country
and not an unfair process of conveyance.
That some unmoneyed colonised
wishing to keep their land
need to sell that land to pay for
legal counsel regretfully recognise
as demonstrating no system is perfect.
Back pioneering legislation that enables
Māori children to heroically advance
educationally socially and economically
by casting off the limitations imposed
on them by their parents who would
naively impede their progress
by having them maintain
their native traditions and language.
Approve clerics' soulful concern
that Māori healers are witchdoctors
and their traditional beliefs ignorant
superstition planted by the Devil.
Back legislation that outlaws native healers
in order to welcome the newly colonised
into the loving bosom of Christianity.
Nod at the Governor's wisdom
in appreciating there will always

be backward-leaning holdouts.
Applaud his enabling handfuls to withdraw
to country back blocks where they
are allowed to retain unwanted land
to scratch a diminished existence.
View this frustratedly as a necessary
life lesson to teach other undecided Māori
the economic social and cultural cost of
withdrawing from the advantages offered
those who embrace the modern world.

With their population rapidly reduced
to under half pre-colonial numbers
accept the widespread judgement that
the Māori are a dying race.
Watch the last living tattooed chiefs
travel to Europe to be admired as
they sip tea with the aristocracy and
wittily and impressively entertain
London's high society madams.
Marvel that the feathers which adorn
those admired chiefs' heads
become a much desired fashion item.
Equally admire the machinery of Empire
as the insatiable market for feathers
leads to the birds being hunted
to extinction which cunningly elevates
the owned feathers' investment value.
Admire the skill of a master painter
who uses a golden palette to capture
that once great people in the last
throes of their exhausted existence.

Applaud these respectful gestures
made by the modern world in generous
genuflection towards a moribund race.

Surprised concede the surviving natives'
zeal to show their loyalty to the Crown
by joining the colonial armies
that sail off to fight in Europe's wars.
Watch them leave their flesh limbs
and swelling bodies strewn across
bloodied wet mud and fly-blown sand.
Acknowledge the war effort is a maw
into which resources must be poured
to defend the new nation against
the glorious Monarch's menacing enemies.
Approve the legislated procurement
of key militarily positioned land
still in tribal hands to protect
the nation from surprise enemy attack.
That the paperwork is later lost
interrupting the land's return
once war's cessation is signalled
consider regrettable but maintain
it cannot besmirch the reputations
of the duteous administrators who served
the emerging nation for meagre reward.
Agree neither property fortunes nor titles
could ever sufficiently repay them
for the selfless sacrifices they made
to tame this wild land.
Resoundingly support the conviction
that a civil society can only stand

on the foundations of requited law.
Regret that when war's survivors return
not only does their strategically annexed land
remain under the control of the State
those battered soldiers unable to pay
the land taxes that accrued
on the modest plots they still own
while they were fighting overseas
must uncomplainingly allow their land
to be seized by the State and onsold
to pay the amounts outstanding.
Assure them that if they continue to play
their loyal and accommodating part
they will be given an equitable place
at the table being set in this new land.

As the second great war resiles into
the sweaty peace of the Cold War
applaud the nation's move to break
its colonial ties and become independent.
Acclaim the new nation's decision
to keep as its revered Head of State
the Empire's glorious Monarch.
When the Monarch deigns to visit
approve local officials' considerate action
of burning down an unsightly Māori village
so her gaze might not be sullied
as she drives past to attend high tea.
That this village housed descendants
of the Māori chief who gifted land
to the nation's first Governor
to establish the nation's largest city

in which that village stood
regard as one of life's great ironies.
Diplomatically observe past attitudes
can never predict future policies as
a massed occupation of that same iwi's
purloined land becomes the launch pad
for a renewed national commitment
to the agreed principles of the Treaty
signed a century before—
that colonised and colonisers equally
participate in the nation's governance
and live in shared bicultural partnership
in which their mutual rights are protected.
Relax as you cook on the Hawaiian beach
knowing the nation your endeavours
helped found is a new-carved waka paddled
into the future by harmonious peoples
entering calm waters that promise
nothing but smooth sailing ahead.

Request

[For Lawrence Ferlinghetti]

I want every day to start well and end better.
I want to look good in every photo.
I want life to be as simple as when I was a kid.
I want non-stick promises free hair gel and to eat jelly
    babies all day without feeling sick.
I want to get back to being a worthwhile person.
I want what I was promised in the womb.

I want to stop feeling like the world is extruding me
    through a money machine.
I want the flow of petrol to reflect the fact I have a soul.
I want to wake each morning feeling like I really want to
    get out of bed.
I want not to feel a corporation is trying to kidnap me
    each time I walk past a billboard.
I want contentment squirted on everyone's tongue
    before they enter the mall.
I want the wonderful life we all deserve.
I want experts to stop telling me what that is.

I want to live in a world where the hippies won.
I want the border opened between what is and what
    could be.
I want ice cream to ooze out of fire hydrants on extra
    hot days.
I want the homeless to levitate into heaven then return
    as angels to drop compassion on our eyelids.

I want Santa's mustachios this Christmas to be
    rainbow-coloured.
I want talk radio not to think that's a terrible idea.
I want everyone to get serious and start dance lessons.
I want the cowboys to give their lassoes to the Indians.
I want to banish war and replace it with campfire
    sing-alongs.
I want the laughing cynic in my head to close the door
    on its way out.

I want a vaccine that inoculates against stupidity.
I want smoke stacks to stop making excuses.
I want religion to relinquish its mortgage on my soul.
I want science to solve the world's most urgent problems
    by next Tuesday.
I want politicians to stop pushing fear buttons.
I want big business to no longer expect tax breaks just
    for existing.
I want men in power to stop making money via slavery.
I want women in power to stop behaving like men in
    power.
I want equality not just to be a good idea.
I want the poor to be treated as well as rich people's pets.
I want the world to grow up so we can take off the
    training wheels.

I want to be like Barbie and become relevant again.
I want to be in a place that's right for me and not where
    others have decided I should be.
I want my identity to no longer be limited to age race
    sex status income possessions looks and who others
    armed with agendas say I am.

I want everyone to view identity as a weird matrix
    spiralling around a strange attractor.
I want life to shoot into the future like the sherbet that
    tingled in my mouth when I was eight.
I want hair like Einstein Johnny Rotten and Phyllis
    Diller.
I want to become the person I always knew I could be.
I want everything to otherwise stay the same.

I want history to absolve us for doing what we know we
    shouldn't but continue to do anyway.
I want what I admit the world is emotionally intellectually
    spiritually and structurally unable to provide.
I want us to collectively acknowledge that problems are
    complex and so are their solutions.
I want everyone to feel that what they are and do counts.
I want us all to love everyone else.
I want that to include non-humans.
I want objectors to jump on their hobby horses and ride
    off into the sunset.
I want the future to become a place where we might
    actually want to live.
I want to get back to whatever it is we agreed we're here
    to do.
I want the world to do much better.
I want someone to drop me a line if any of this happens.

Acknowledgements

The animal park in Neumünster ...
From *German zoo to feed animals to each other*, originally published by Die Velt, republished by the New Zealand Herald, 16 April, 2020.

Humanity is waging war on nature ...
From a speech given by the UN's Secretary General in December 2020, widely reported by world media.

The Walton family benefited enormously ...
From *How the world's richest families teach their kids to hoard wealth and create billionaire 'dynasties'*, written by Alex Turner-Cohen, published by the New Zealand Herald, 26 June, 2021.

The RAND Corporation ...
From *The Elites Are Fighting a Vicious Class War All the Time*, an interview with Noam Chomsky by Ana Kasparian and Nando Vila, published by Jacobin, 10 June 2021: jacobinmag.com. This interview is also on Noam Chomsky's website: chomsky.info/20210610/

Rodney: "The peasants are revolting" ...
From the comic strip *The Wizard of Id* by Brant Parker and Johnny Hart, which debuted in The Philadelphia Inquirer, 8 November 1964.

"These things are because we're living ..."
From interviews of Papua New Guineans, researched and written by Lorelle Tekopiri Yakam and Kylie McKenna, originally published in a blog, *Climate change: a sign of the 'End Times'*, republished online on DEVPOLICYBLOG, 18 September 2020.

"I believe we need Christian men and women ..."
Quoted in a news article, *Mississippi secretary of state calls for more Christians to be elected because 'the end times' are here*, written by David Babash, published by Alternet, 7 May 2021.

In the 1950s about 75 percent of Americans trusted ...
Statistics quoted in *Blowing up the billionaires' con that's strangling America*, written by Thom Hartmann, published online on The Hartmann Report, 22 May 2121: hartmannreport.com.

80% of Australians and 83% of New Zealanders ...
From research carried out by Shaun Goldfinch of Curtin Universty, Robin Gauld of the University of Otago, and Ross Taplin of Curtin University, that is included in an article, *How Covid-19 changed public trust in governments in New Zealand and Australia*, published by The Spinoff, 12 Februrary 2021.

About the author

Keith Hill is a writer and filmmaker living in Aotearoa New Zealand who works across a range of creative and educational fields.

His interest in facilitating others' creativity led him to co-found Rattle Records in 1991. For nineteen years he worked with some of the country's leading classical, jazz and world musicians, including Hirini Melbourne, Richard Nuns, Michael Houston, Phil Dadson, NZ String Quartet and John Psathas. Over this period Rattle won Best Classical Album in the NZ Music Awards five times and the ground-breaking *Te Ku Te Whe* went platinum. Keith videotaped many recording sessions to archive the musicians' work. *Persuading the Baby to Float* (2012), documenting setting Bill Manhire's poetry to music, premiered in the Wellington International Film Festival.

Since the 1980s Keith has worked in New Zealand's screen industries. Short films he produced premiered in many local and international film festivals. After a year as visiting lecturer in film at Ilam School of Fine Arts, University of Canterbury, Keith joined the Waikato Institute of Technology, where he subsequently became head of the Moving Image programme. His feature film, *This is Not a Love Story*, won Best Screenplay Award in a US indie film festival.

In his writing, Keith is drawn to the ways culture, history, literature and spirituality intersect. Travel to India resulted in Keith working on translations of India's greatest mystic poets, Mirabai and Kabir, and a poetic version of the *Bhagavad Gita*. Other books examine how our ideas about morality, metaphysics and spirituality have radically changed since the rise of the sciences. His books won NZ's Ashton Wylie Awards three times, and were finalists many other years.

This book has risen out of frustration that we are spiritual beings who have a duty of care towards this planet and those we share it with, but we are instead decimating it.

To the reader

Small presses rely on the support of readers to tell others about the books they enjoy. To support this book and its author, we ask you to consider placing a review on the site where you bought it. For more on Keith Hill and his books see: www.keithhillauthor.com. Keith Hill's other books include:

The Ecstasy of Cabeza de Vaca

""A tour de force. Hill's humanizing of de Vaca is the ingredient that makes it so moving and once taken up, impossible to put down."
— Alistair Paterson
"An extraordinary effort of imagination. In New Zealand literature there's no one quite like Keith Hill, and certainly no long poem like this one." — Roger Horrocks

The Lounging Lizard Poet of the Floating World

"Is this book ticking?" — Receptionist, Ministry for Culture
"Offers an insightful and kaleidoscopic dissection of the modern world. It's an entertainingly timely invitation to wake up." — Hugh Major, author of *The Lantern in the Skull*

Interpretations of Desire: Mystical love poems by Ibn 'Arabi

"The *Tarjumán al-Aswáq* is one of the greatest works of Islamic mystical poetry. Keith Hill's artful and beautiful renditions will bring Ibn 'Arabi's neglected masterpiece to a new readership."
— Nile Green, author of *Sufism: A Global History*

I Cannot Live Without You: Poems of Mirabai and Kabir

"Reminded me it's been an eternity since I was hungry for God. This book will renew your hunger for your sacred flame." — Judith Hoch PhD, author of *Prophecy on the River*

Psalms of Exile and Return
"In a time that seems spiritually dry for so many, this book of psalms is water in the desert. They challenge, terrify, comfort, and call us to a deep humanity." — Allan Jones, Dean Emeritus, Grace Cathedral, San Francisco

The Bhagavad Gita: A new poetic translation
"An enthralling new rendering, which balances spiritual insight, poetic power and philosophic accuracy." — Peter Calvert, co-author of *The Kosmic Web*

Puck of the Starways
"It's a masterpiece. But I have no idea how you could market it."
— John Psathas

The God Revolution
Winner, Best Book Ashton Wylie Awards 2011
"Hill's exposition is a fine example of scrupulously rigorous scholarship—it is remarkable how much ground is covered within his brief historical survey. In addition, he discusses a wide range of academically abstruse subjects in consistently lucid, nontechnical prose."
— Kirkus Review
"A scholarly yet accessible book ... Books of this calibre, written and published here in New Zealand, are a rare phenomena. This deserves to be read by all those who care about ideas, the trajectory of civilization and its future form." — Peter Dornauf, www.eyecontact.com

CPSIA information can be obtained
at www.ICGtesting.com
Printed in the USA
BVHW061746300422
635802BV00006B/335

9 781991 157072